Written by Raphaelle Brice
Illustrated by Aline Riquier

Specialist adviser: Robert Press,
Botanical Consultant

ISBN 1 85103 016 6
First published 1986 in the United Kingdom by
Moonlight Publishing Ltd.,
131 Kensington Church Street, London W8

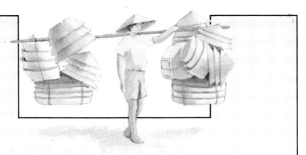

The Story of a Grain of Rice

You know how tiny each grain is,
    yet rice is the most important
                food on earth...

Rice for every meal.

If you lived in China, India, Vietnam, or any other country in Asia, you would eat rice every day: usually just plain rice for your midday meal, while in the evening, as well as rice, you might have meat, fish or vegetables.

How do you eat rice?

It always has to be cooked, then you can eat it hot or cold, with a little salt, or with sugar and milk, as rice pudding. Or it can be made into cakes. Perhaps you have puffed rice to eat for breakfast, as a cereal with milk and sugar. In some countries, they eat salted puffed rice instead of crisps or peanuts.

Look at all the different kinds of rice dish.
Do you know any others?

Rice grows with its feet in the water, its head in the warm sun.

The stalk is hollow, so that the plant can suck up water to feed itself.

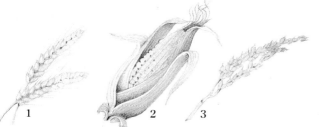

1   2   3

Rice is a cereal, like wheat (1), maize (2), or oats (3). Cereals are plants which give seed which can be eaten, or ground up to make flour.

We only eat the kernel of the rice plant seed.

Here is a grain of rice which has been cut in half. You can see that as it grows, it is protected by a fine envelope: the husk. The grain contains the seed of a new rice plant. It would grow if you planted it. The plants grow in clumps, each producing several ears.

Some rice has long grains, other kinds have round grains.

People have been eating rice for thousands of years.

The Chinese and Indians first learnt to cultivate it, about 7,000 years ago. The way rice is grown has hardly changed since then.

Rice has travelled the world.

Merchants, soldiers and sailors carried rice from Asia to the countries they visited. The Arabs introduced rice to Europe only about 800 years ago.

How did rice get to America?

About 350 years ago, a ship sailing from Madagascar sank off the American coast. The captain thanked his rescuers by giving them sacks of rice.

Today, more than half the people in the world eat rice.
Nine out of every ten sacks of rice contain rice grown in Asia. The Asians eat almost all of the rice they grow.

In the United States, too, a huge amount of rice is grown. A lot of it is sold to the rest of the world. A little is grown in southern Europe, but most of the rice eaten in Europe is bought from America.

Buffaloes do not mind the water, and hardly sink down in the muddy soil. But they get tired very easily, and can only work for 2 hours a day.

This is Asia, the continent of rice.

To grow the rice, the peasants have to flood the fields: these are called **paddy-fields.** Ditches carry the water from

This tractor, with its hollow wheels, doesn't sink down into the mud.

the river to the fields. Little barriers, or bunds, can be lifted or lowered to let the water flow, or to stop it. When the soil is muddy, it is raked to get out the lumps, and then flattened with a harrow. The rice is first sown in a separate field: the seed-bed.

The peasants sometimes scoop water from one field to another with baskets.

15

After a month has passed, the little rice plants have grown. They are now too close together, and need to be replanted in the paddy-fields.

Bent double, up to their ankles in mud, women plant out the rice.
The plants are laid out in straight lines, with plenty of space between them so that they have room to grow. The water discourages the weeds. Depending on how hot the country is, the rice takes three to six months to grow. While they are waiting, the peasants keep an eye on the water level, and make sure the birds don't eat the crop.

Did you know there are thousands of varieties of rice in the world?

Paddy-fields are a paradise for birds.
Herons, long-legged cranes, wild
ducks, all find their food in the
flooded fields. They eat frogs, fish,
snails, algae, and, of course, the
ripening grains of rice.

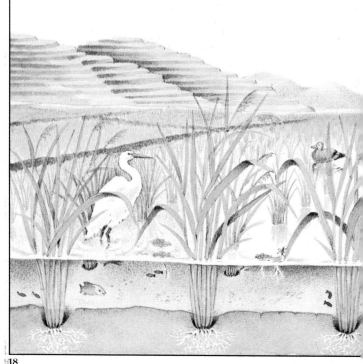

But these damp places also attract mosquitoes, which attack people, and a host of other insects, which attack the rice. Some bugs and butterfly larvae are the most harmful, gnawing away at the stalks and leaves.

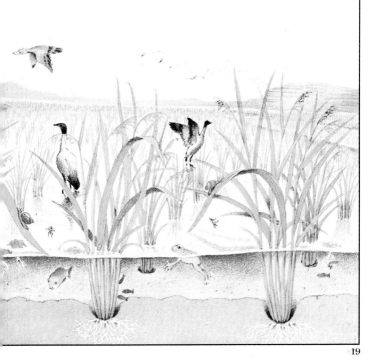

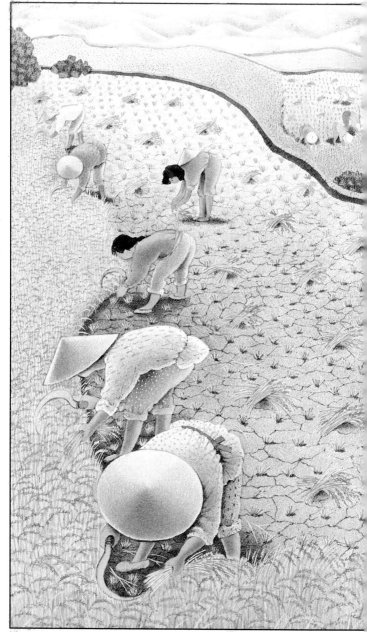

The rice has turned gold, the heads are heavy with grain.

The rice finishes ripening with 'dry feet'.
Just before the harvest, the peasants lift the barriers and let all the water drain out back into the river. The mud in the paddy-fields grows hard and dry.

The harvest.

From dawn to dusk, the rice-growers cut the stalks with their sickles. They tie the plants in bunches, then carry them back to the village, either by cart, or on long poles.
What happens to the rice then?

The rice is beaten and dried

The peasants beat the stalks against a stone roller to shake off the grain. The stalks are arranged into straw-stacks. The grains of rice are spread out, and turned by foot or with a rake so that they dry evenly in the sun.

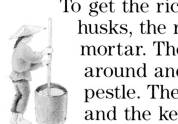

The rice is husked.

To get the rice grains out of the husks, the rice is put into a mortar. Then it is pushed around and crushed with a pestle. The husks split open, and the kernels are set free.

The rice is winnowed.

Set out on huge bamboo platters, and beaten with rods, the husks fly away in the wind, while the heavier grain stays at the bottom.

A rice granary

In some countries, and especially in islands like the Philippines, the peasants grow rice right up the mountains. They dig terraces into the mountain-side, shoring them up with

little walls of stone. Water from springs runs from terrace to terrace, flowing down in special channels from the tops of the mountains.

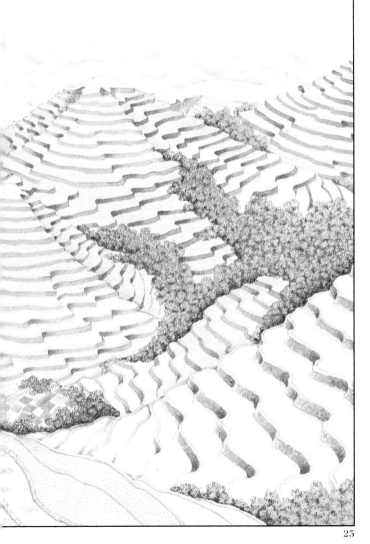

The land of machines

In the United States, rice is grown in huge fields.

Nearly everything is done with enormous machines.

The fields are raked and harrowed by tractor. So as to flood the fields, channels are dug by machine, and then filled from a big pipe.

The rice germinates
in sacks soaking
in water.

Airplanes sow the rice. There is no
replanting.

Combine-harvesters gather in the ripe
plants and automatically separate the
straw from the grain. In the factory,
the rice is dried, husked and put into
packets.

Is rice only useful as food?

Rice-straw can be used to make baskets, mats, and large flat hats to shelter the rice-growers from the sun.

The rice husks are used to feed animals, or to make fertiliser which is spread on the fields to put the goodness back into the soil ready for next year's crop.

Ground rice is made into flour which the Asians use for cooking to make cakes and noodles. They make wine and beer from rice as well. It can even be used as face-powder.

Is rice white or brown?

When it has still got its husks on, rice is called paddy. When it has been husked, it is brown; this is sometimes called whole-grain rice. In order to make white rice, the grain must be put into special machines which remove the last little bits of husk.

| Paddy rice | Brown rice | White rice |

Rice festivals

In Asia, the peasants organise festivals to ask the gods and the spirits of the rice for a good harvest.

A Taradja granary

Every year, in the mountains of the island of Sulawesi, the Taradja people sing and dance all night long, miming the work that is done in the fields. Every family has a granary beside its house. It is built on stilts, with a curved roof to protect the crop from the rain.

The peasants on the island of Bali offer rice cakes to the gods of their temples.

A shower of rice

In Europe and America, rice is thrown over newly married couples to wish them good luck.

How is rice cooked in Asia?

Most often it is steamed. Water is put in a pot and set to boil, and a cloud of steam rises to cook the rice which has been put in a basket on top.

This Jamaican woman is cooking rice in the same way that you probably do it at home: in a big saucepan of boiling water.

In India and Pakistan rice is eaten mixed with meat and vegetables in hot, spicy sauces.

Rice is very good for you. It is full of vitamins, which help your body to grow and to keep healthy. In countries where rice is eaten every day, it is as important, and as much enjoyed, as bread or potatoes for you.

For 4 people you will need

400g of rice

3 eggs

Salt, pepper

Butter

A thick slice of ham

Cooked peas

Cantonese rice: a Chinese recipe for the whole family

Cook the rice in boiling salted water. Cut the ham into small cubes. Use the eggs to make an omelette, then cut the omelette into strips. Drain the rice and mix it with the ham, eggs, and peas. Salt, pepper and dot with butter, and leave it to stand for a few minutes in a hot buttered dish.

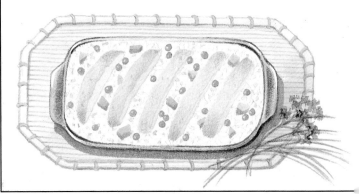

How do you eat with chopsticks?

Choose rice which sticks together slightly after it's been cooked: it's easier to pick up. Look at the pictures at the top: the lower chopstick is held between your thumb and third finger, and doesn't move. The top chopstick, held between the tip of your thumb and your index finger, picks the rice up by pressing it against the lower chopstick. Hold the bowl close to your face, and good luck!

Index